AF887697

Leaping Learners
Education, LLC

For more information and resources visit us at:
www.leapinglearnersed.com

Special thanks to our reading level consultant
FRANCINE DEL VECCHIO, Ed.D.
Director of the Literacy Center
Professor of Education
Caldwell University

ISBN
978-1-948569-04-0

Dear Parents and Guardians,

Thank you for purchasing a *Fay Learns About* series book! After teaching students from kindergarten to second grade for more than seven years, I became frustrated by the lack of engaging books my students could read independently. To help my students engage with nonfiction topics, my wife and I decided to write nonfiction books for children. We hope to inspire young children to learn about the natural world.

Here at Leaping Learners, LLC, we have three main goals:

1. Spark young readers' curiosity about the natural world
2. Develop critical independent reading skills at an early age
3. Develop reading comprehension skills before and after reading

We hope your child enjoys learning with Fay. If you or your children are interested in a topic we have not written about yet, send us an email with your topic, and maybe your book will be next!

Thank you,

Sean Bulger, Ed.M

www.leapinglearnersed.com

Reading Suggestions:

Before reading this book, encourage your children to do a "picture walk," where they skim through the book and look at the pictures to help them think about what they already know about the topic. Thinking about what they already know helps children get excited about learning more facts and begin reading with some confidence.

Preview any new vocabulary words with your child. Key vocabulary words are found on the last few pages of the book. Have your children use each new phrase in their own words to see if they understand the definition.

After previewing the book, encourage your child to read the book independently more than once. After they have read it, ask them specific questions related to the information in the book. Encourage them to go back and reread the relevant section in the book to retrieve the answer in case they forgot the facts.

Finally, see if your child can complete the reading comprehension exercises at the end of the book to strengthen their understanding of the topic!

This book is best for ages 7-8
but. . .
Please be mindful that reading levels are a guide and vary depending on a child's skills and needs.

Fay Learns About . . . Sea Stars

Written by Sean and Anicia Bulger

Table of Contents

Hi! My name is Fay. I love to discover and learn new things. In this book, we will learn about sea stars. Let's go!

Introduction

Have you ever seen a sea star?

These interesting animals have a special body to survive in the ocean.

Body

Sea stars are spiny skinned animals.

Sea stars come in many different colors and have arms called rays.

Ray

Sea Stars have five or more rays that point out in every direction from the center of their small round body.

The rays have tiny **suction cups** underneath them that act as tubed feet. These allow them to grip things, which helps them move around and get food.

How do sea stars eat?

stomach

Sea stars have two stomachs. One is on the outside of the body and the other is found inside the body. The outer stomach eats the food and transfers it to the inner stomach, which breaks down the food.

Sea stars can **regenerate** their arms if they fall off! This helps them get away if a predator catches them.

TRICKY WORD
CHUNK IT OUT
RE-GEN-ER-ATE
Regeneration

Sea Stars come in many different colors. These colors help them **camouflage** in their habitat to stay safe.

Habitat

Sea stars can be found in all the world's oceans. They live in tide pools, coral reefs, and even on the ocean floor!

Diet

A sea star's diet mostly includes **bivalves**. Some of a sea star's favorite bivalves are mussels, clams, and oysters.

Mussels

Tircky word
Chunk it out
bi-valve

Clams

Oysters

Sea stars eat by **prying** open the shells of their prey with the suction cups on their arms. Then the sea star pushes its stomach out of its mouth into the shell to eat the prey!

Predators

Crabs

Sea turtles

A sea star has many predators. Some include sea turtles, fish, sea otters, and crabs. These predators hunt for sea stars on rocks and coral.

Sea otters

Fish

Fun Facts

Sea stars can have up to 50 arms!

Sea stars don't have a brain!

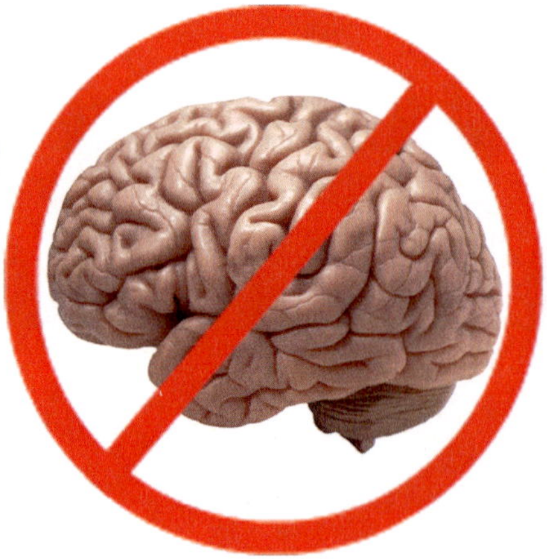

Sea stars don't have blood. They pump water through their bodies.

Glossary

A glossary tells the reader the meaning of important words.

Suction cups – Cup like part of an animal, usually on their arms or tentacles. Suction cups grip things using pressure.

Camouflage – To blend into the surrounding environment.

Bivalve – Animal with no spine and is covered by two shells

Pry – To pull something open using force

Regenerate – To regrow a body part

Draw a picture of a sea star

Choose 3 words from the glossary and write a sentence for each one.

1.____

2.____

3.____

Quiz

1. What helps sea stars move and eat?
a. With their suction cups
b. With their flippers
c. With their feet

2. Which animal eats sea stars?
a. Snails
b. Sea turtles
c. Clams

3. How can sea stars escape danger?
a. Swimming away
b. Losing an arm
c. Biting animals

Common core standards:
RI. 2.1 - Questions 1, 2, 3

4. How many stomachs do sea stars have?
a. 1
b. 2
c. 3

5. What do sea stars eat?
a. Plants
b. Crabs
c. Clams

6. In which section do you learn about prey?
a. Body
b. Habitat
c. Diet

Common core standards:
RI. 2.1 - Questions 4, 5
RI. 2. 2 - Question 6

Answer the following questions using information from the text:

1. How do sea stars eat their food?

____

____

____

____

____

2. How do sea stars stay safe?

____

____

____

____

____

3. Where do predators hunt for sea stars?

4. What new information did you learn?

Want to learn about rainforest animals? Check out the "Matt Learns About..." series!

Great for emerging readers ages 6-8

Want to learn about colors? Check out the "Clayton Teaches you About..." Series!

Great for early readers ages 4-6

Want to learn about Farm Animals? Check out the "Katie Teaches you About..." Series!

Great for early readers ages 4-6

Made in United States
Orlando, FL
02 November 2026

90104226R00024